For Todd —
Wishing you great adventure
and joy.
Fondly,
Carolyn Lesser
1987

THE GOODNIGHT CIRCLE

by Carolyn Lesser

Illustrated by
Lorinda Bryan Cauley

Harcourt Brace Jovanovich, Publishers
San Diego   New York   London

Requests for permission to make copies of any part of the work should
be mailed to: Permissions, Harcourt Brace Jovanovich, Publishers,
Orlando, Florida 32887

Library of Congress Cataloging in Publication Data

Lesser, Carolyn.
The goodnight circle.
Summary: Describes the activities of a variety of
animals from sunset to sunrise.
1. Animals — Juvenile literature.   2. Nocturnal animals
— Juvenile literature.   3. Biological rhythms — Juvenile
literature. [1. Nocturnal animals.   2. Animals]
I. Cauley, Lorinda Bryan, ill.   II. Title.
QL49.L384 1984   591.51   84-4501
ISBN 0-15-232158-6

Printed in the United States of America

Designed by Mark Likgalter

B   C   D   E

For my parents and for Harry

—C. L.

For the Fitzpatrick family, with affection

—L. B. C.

The goodnight circle begins, as the setting sun turns the sky
orange. The warm spring air cools. Mother deer nudges her
fawns onto a bed of soft, fragrant pine needles. The fawns
nestle close to mother deer, as she lovingly licks their fur.
Good night, pretty deer.

It is scold and chatter, all day, as the noisy squirrels show
their ten babies where to find nuts. At last evening comes.
All the babies are safe in the cozy nest, high in the hollow
tree. Mother squirrel ducks into the hole and snuggles with
her babies. Good night, busy squirrels.

A face with a shiny black nose, bright beady eyes, and pointy ears as orange as flame, peeks around a big rock. Eyes darting, mother fox is taking no chances. Hide and chase—slink and pounce—dodge and dart, are today's lessons. Slyness and cunning—trickery and winning, will keep her babies alive.

The sky is becoming darker, as mother fox hurries her pups
into the den. They crawl through winding tunnels, deep
inside the steep bank of the hill, to safety. Mother fox
follows, and curls up. The pups press close to her, their
fluffy tails like blankets around them. Good night,
smart little foxes.

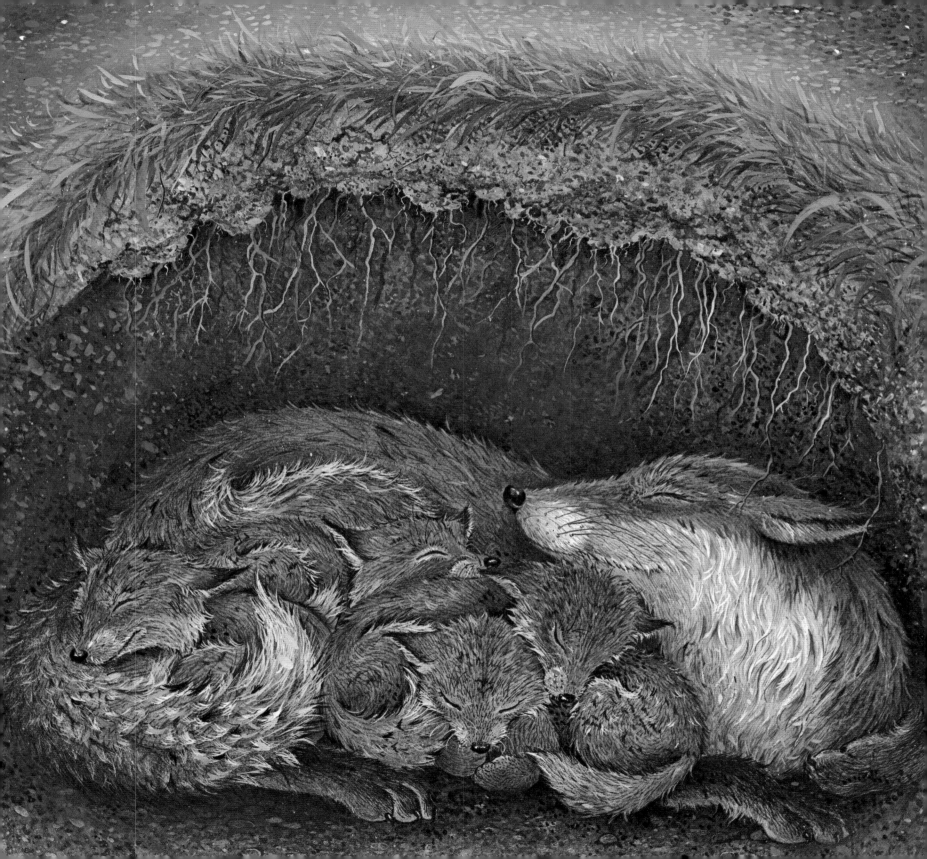

The huge black crows, in their nests of sticks in the tall pine tree, gaze down on the pond. They look for food every minute for their screeching, scrawny, fast-growing babies.

They see a box turtle, his red eyes glistening in the setting sun, his brown and gold shell the color of the bushes. Night is coming, so the turtle carefully crawls deep into a thicket of thorny brambles. He pulls his head in, and closes the hinged bottom of his shell. Good night, sleeping rock.

The crows see the water striders swimming on the top of the
sparkling pond. The tiny insects pull their thread-thin legs
like oars as they skim across the pond. They are too small to
feed crow babies. Now they find their beds in the grasses
growing in the pond. Good night, tiny rowers.

Suddenly father crow plunges down at the brown and gold garter snake, but the snake quickly glides beneath a rock. Her babies would be a fine dinner for young crows. Father crow keeps his beady black eyes on that rock, but mother snake will not come out.   Good night, slithery snakes.

Worms will have to do for the hungry crows. As father crow pulls the worms from the soft earth, he disturbs the bullfrog. The frog grunts to the other frogs in the pond. It is time for him to find his squishy mud bed. Good night, fat frog.

As the last
shred of gray leaves the sky,
the black-and-white woodpecker
clings to the bark of the hickory tree.
He pokes his beak into the bark, eating
the insects that live there. Only the
bright red spot on his head can be
seen in the fading daylight. With
a flap of wings, he is off to his
nest. Good night, small
driller.

For a moment, the forest is silent as the day animals sleep. Stars fill the sky. Moonlight makes the pond shimmer like silver. The goodnight circle grows.

Luna moths, big as saucers, sail through the night, like pale green ships. The luna moths will fly all night, resting on branches as their wings tire. Good morning, beautiful moths!

Night wakes the owl. He glides from the gnarled
branch of the oak tree, his wings flapping silently.
The owl circles the forest, his keen eyes searching the
ground for food, his sharp ears listening for any tiny sound.
Good morning, hungry owl!

Tiny gray mice scamper from log to rock
to toadstool, always hiding on this moonlit night.
They make no sound. Mother mouse teaches them about
the big silent owl. Quietly they nibble nuts and berries.
Good morning, whispery mice!

The spring peepers break the moment of silence between day and night. The tiny frogs, clinging to stalks of plants, sing songs of spring to the moon and the stars. Good morning, cheerful singers!

Mother opossum, trudging toward the pond, hears something
scary. She runs up the nearest tree, her babies clutching her tail.
Food must wait tonight. The most important lesson is—run from
danger. Good morning, careful opossums!

Trees crash. The forest floor shakes.
The night animals tremble.

The beavers, young and old, are at work! They scurry to the
fallen trees, their big, flat tails dragging over the ground.
They strip tender young limbs from the trees. Carrying the
branches in their teeth, they return to the pond to dive
underwater and stick branches in the mud.

All night long, the beavers in the pond work together to find
food and keep the dam from breaking. Good morning,
hardworking beavers!

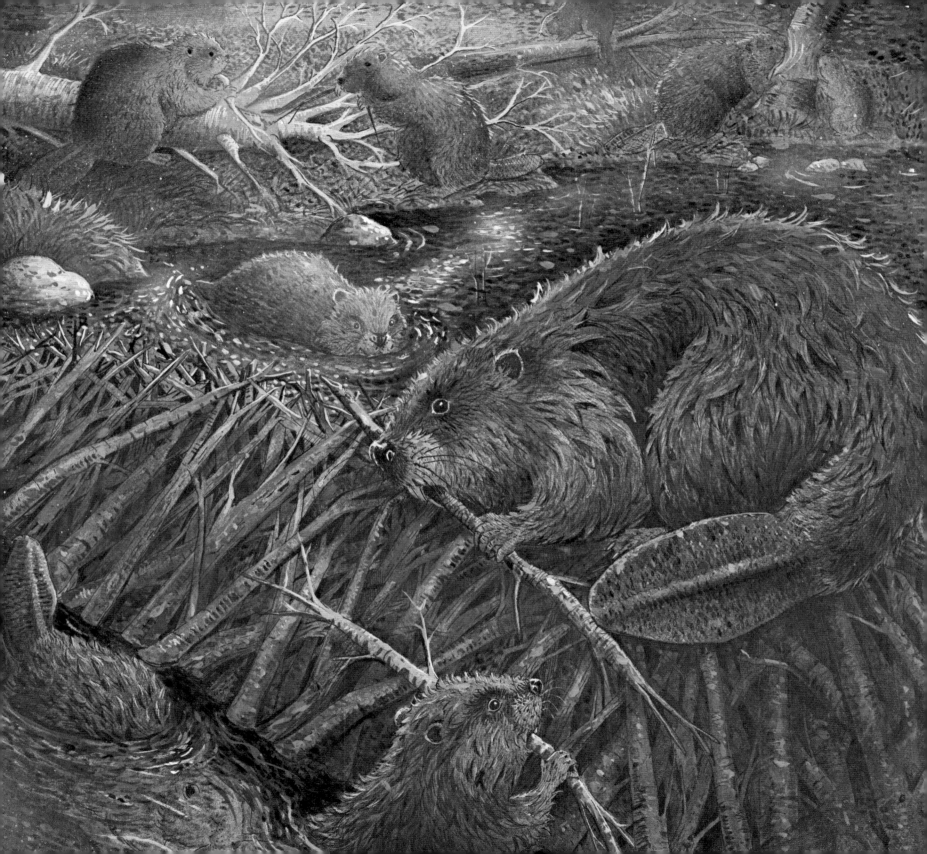

The gray-green pickerel swims past the beavers to the grasses
at the edge of the pond. He nips at the water striders hiding
there. He will eat now, and sleep behind a big rock when
daylight comes. Good morning, swimming fish!

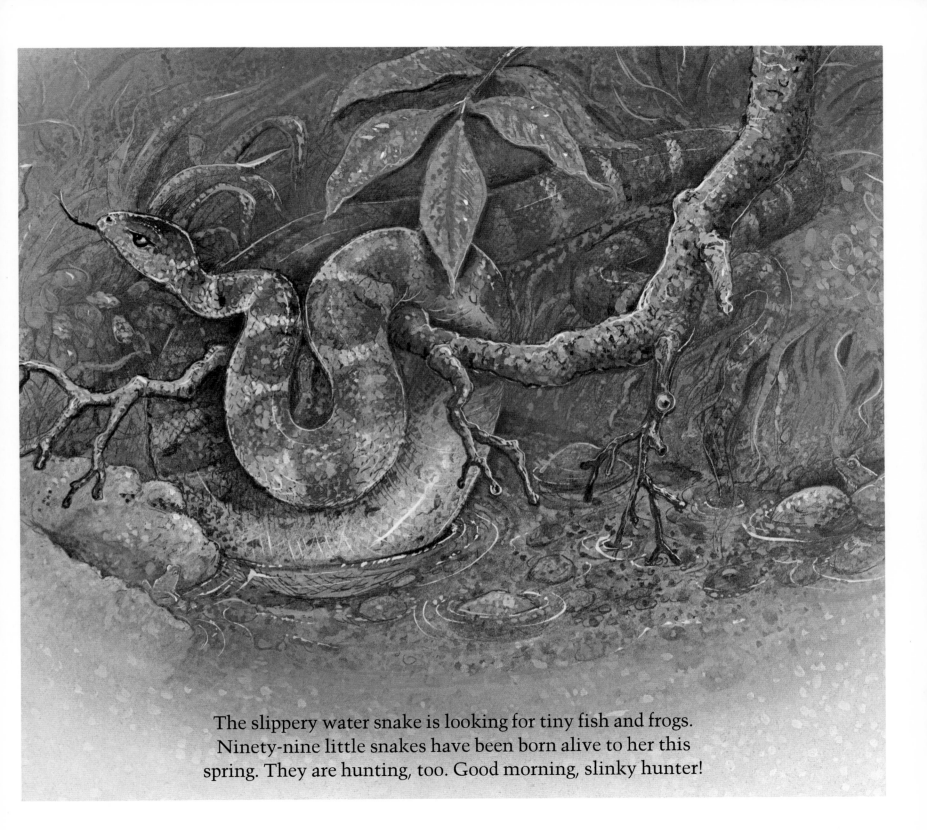

The slippery water snake is looking for tiny fish and frogs. Ninety-nine little snakes have been born alive to her this spring. They are hunting, too. Good morning, slinky hunter!

The goodnight circle is closing, as a masked face peeks around the big oak tree. Suddenly, pop! pop! pop! Three small masked faces appear. Mother raccoon leads her bandits to the edge of the pond.

She slaps a sparkling crayfish onto the bank. Her babies grab it. Mother raccoon shoves one baby into the water and nudges him until he begins to slap the water. Out flips a tiny crayfish. The baby pounces on it. By the gray of early dawn, all three baby raccoons are sloshing the water happily. Good morning, little fishermen!

The night animals are busy, teaching and learning, eating and playing. Now, as soft, pink sunrise comes, they find their beds. Dewdrops glisten on the unfurling leaves of the wood violets. The night animals sleep.

For a moment, the forest is silent. It is the moment
between night and day.

Mother deer nuzzles her sleepy fawns awake. They shake themselves and stand like statues, gazing at the misty-morning forest. The crunch of hooves on the forest floor shatters the silence, as they bound off to search for breakfast.

A new day begins. The goodnight circle is complete.